Fractions, Division & Multiplication

2nd Grade Math Workbook Series Vol 3

SPEEDY
PUBLISHING

Speedy Publishing LLC
40 E. Main St. #1156
Newark, DE 19711
www.speedypublishing.com

Multiplication

Name:_____

1. 12 x 10 =

4. 25 x 48 =

2. 73 x 62 =

5. 94 x 57 =

3. 27 x 23 =

6. 53 x 45 =

Name:_____

1. 83 x 47 =   **4.** 22 x 80 =

2. 84 x 21 =   **5.** 31 x 96 =

3. 31 x 43 =   **6.** 72 x 79 =

Name:_____

1. 81 x 73 =   **4.** 16 x 24 =

2. 30 x 80 =   **5.** 25 x 14 =

3. 53 x 73 =   **6.** 58 x 14 =

4

Name:_____

1. 33 x 40 =   **4.** 41 x 65 =

2. 33 x 64 =   **5.** 20 x 24 =

3. 43 x 43 =   **6.** 26 x 20 =

Name:_____

1. 97 x 85 =   **4.** 88 x 100 =

2. 85 x 55 =   **5.** 12 x 98 =

3. 53 x 14 =   **6.** 59 x 99 =

6

Name:_____

1. 51 x 15 =   **4.** 77 x 39 =

2. 88 x 65 =   **5.** 86 x 49 =

3. 98 x 65 =   **6.** 37 x 33 =

7

Name:_____

1.   64   x   48   =

4.   50   x   24   =

2.   25   x   26   =

5.   37   x   55   =

3.   62   x   22   =

6.   10   x   69   =

8

Name:_____

1. 49 x 61 =   4. 20 x 62 =

2. 59 x 56 =   5. 93 x 16 =

3. 69 x 82 =   6. 63 x 66 =

9

Name:_____

1. 79 x 91 =   **4.** 10 x 57 =

2. 79 x 47 =   **5.** 38 x 62 =

3. 62 x 68 =   **6.** 91 x 13 =

10

Name:_____

1. 11 x 32 =   **4.** 44 x 50 =

2. 37 x 77 =   **5.** 90 x 93 =

3. 74 x 34 =   **6.** 45 x 85 =

Name:_____

1. 65 x 67 =   **4.** 41 x 65 =

2. 80 x 63 =   **5.** 98 x 66 =

3. 96 x 42 =   **6.** 43 x 75 =

Division

Name:_____

1. $580 \div 20 =$ **4.** $672 \div 28 =$

2. $1722 \div 42 =$ **5.** $624 \div 48 =$

3. $680 \div 17 =$ **6.** $2115 \div 47 =$

Name:_____

1.   546 ÷ 39 =   **4.**   288 ÷ 16 =

2.   364 ÷ 14 =   **5.**   1116 ÷ 36 =

3.   630 ÷ 35 =   **6.**   494 ÷ 19 =

3

Name:_____

1.   392   ÷   14   =   **4.**   1350   ÷   27   =

2.   1204   ÷   28   =   **5.**   950   ÷   38   =

3.   1221   ÷   37   =   **6.**   143   ÷   13   =

4

Name:_____

1. $1080 \div 36 =$ 4. $336 \div 12 =$

2. $812 \div 29 =$ 5. $1140 \div 38 =$

3. $286 \div 26 =$ 6. $560 \div 35 =$

5

Name:_____

1. 2024 ÷ 44 =   **4.** 308 ÷ 14 =

2. 1634 ÷ 43 =   **5.** 810 ÷ 18 =

3. 140 ÷ 10 =   **6.** 320 ÷ 10 =

6

Name:_____

1.   1550 ÷ 31 =

4.   378 ÷ 27 =

2.   341 ÷ 31 =

5.   1496 ÷ 34 =

3.   714 ÷ 21 =

6.   940 ÷ 47 =

7

1. $319 \div 29 =$ **4.** $1148 \div 41 =$

2. $120 \div 12 =$ **5.** $644 \div 23 =$

3. $640 \div 32 =$ **6.** $1104 \div 48 =$

8

Name:_____

1. $1485 \div 33 =$ **4.** $231 \div 11 =$

2. $714 \div 34 =$ **5.** $902 \div 41 =$

3. $680 \div 20 =$ **6.** $1000 \div 20 =$

9

Name:_____

1. 988 ÷ 38 =   **4.** 450 ÷ 45 =

2. 1150 ÷ 50 =   **5.** 310 ÷ 31 =

3. 1872 ÷ 39 =   **6.** 1216 ÷ 38 =

10

Name:_____

1.   594 ÷ 18 =   **4.**   1040 ÷ 26 =

2.   416 ÷ 16 =   **5.**   800 ÷ 25 =

3.   775 ÷ 25 =   **6.**   286 ÷ 22 =

Name:_____

1.   1036   ÷   37   =   **4.**   1419   ÷   43   =

2.   435   ÷   29   =   **5.**   1505   ÷   43   =

3.   1247   ÷   29   =   **6.**   931   ÷   19   =

Fractions

(Multiplication and Division)

Name:_____

1. $\dfrac{8}{9}$ × $\dfrac{5}{10}$ =

3. $\dfrac{2}{10}$ × $\dfrac{7}{10}$ =

2. $\dfrac{4}{8}$ × $\dfrac{9}{9}$ =

4. $\dfrac{3}{10}$ × $\dfrac{8}{8}$ =

2

Name:_____

1. $\dfrac{6}{9}$ × $\dfrac{7}{9}$ =   3. $\dfrac{3}{8}$ × $\dfrac{5}{8}$ =

2. $\dfrac{7}{7}$ × $\dfrac{1}{10}$ =   4. $\dfrac{6}{8}$ × $\dfrac{7}{8}$ =

Name:_____

1. $\dfrac{1}{9} \times \dfrac{2}{9} =$ 3. $\dfrac{6}{10} \times \dfrac{2}{8} =$

2. $\dfrac{3}{9} \times \dfrac{8}{10} =$ 4. $\dfrac{4}{10} \times \dfrac{1}{8} =$

4

Name:_____

1. $\dfrac{5}{9} \times \dfrac{4}{9} =$ 3. $\dfrac{7}{9} \times \dfrac{6}{9} =$

2. $\dfrac{4}{9} \times \dfrac{8}{9} =$ 4. $\dfrac{8}{8} \times \dfrac{3}{10} =$

5

Name:_____

1. $\dfrac{7}{10} \times \dfrac{2}{10} =$ 3. $\dfrac{5}{10} \times \dfrac{8}{9} =$

2. $\dfrac{9}{9} \times \dfrac{4}{8} =$ 4. $\dfrac{7}{7} \times \dfrac{8}{10} =$

6

1. $\dfrac{1}{10}$ × $\dfrac{6}{10}$ =   3. $\dfrac{5}{8}$ × $\dfrac{4}{10}$ =

2. $\dfrac{3}{8}$ × $\dfrac{2}{8}$ =   4. $\dfrac{6}{8}$ × $\dfrac{1}{8}$ =

Name:_____

1. $\dfrac{7}{8} \div \dfrac{5}{9} =$ 3. $\dfrac{2}{9} \div \dfrac{6}{9} =$

2. $\dfrac{1}{9} \div \dfrac{4}{9} =$ 4. $\dfrac{7}{9} \div \dfrac{3}{9} =$

8

Name:_____

1. $\dfrac{6}{9} \div \dfrac{2}{9} =$ 3. $\dfrac{3}{10} \div \dfrac{7}{8} =$

2. $\dfrac{8}{8} \div \dfrac{1}{9} =$ 4. $\dfrac{7}{10} \div \dfrac{6}{8} =$

9

Name:_____

1. $\dfrac{2}{10} \div \dfrac{5}{8} =$ 3. $\dfrac{4}{8} \div \dfrac{1}{10} =$

2. $\dfrac{9}{9} \div \dfrac{3}{8} =$ 4. $\dfrac{5}{10} \div \dfrac{7}{7} =$

Name:_____

1. $\dfrac{8}{9} \div \dfrac{7}{7} =$ 3. $\dfrac{4}{10} \div \dfrac{5}{8} =$

2. $\dfrac{3}{8} \div \dfrac{2}{8} =$ 4. $\dfrac{1}{8} \div \dfrac{6}{8} =$

Name:_____

1. $\dfrac{5}{9} \div \dfrac{7}{8} =$ 3. $\dfrac{3}{9} \div \dfrac{2}{9} =$

2. $\dfrac{4}{9} \div \dfrac{1}{9} =$ 4. $\dfrac{3}{10} \div \dfrac{7}{10} =$

ANSWERS

1. 120	**1.** 5913	**1.** 8245	**1.** 3072	**1.** 7189	**1.** 4355
2. 4526	**2.** 2400	**2.** 4675	**2.** 650	**2.** 3713	**2.** 5040
3. 621	**3.** 3869	**3.** 742	**3.** 1364	**3.** 4216	**3.** 4032
4. 1200	**4.** 384	**4.** 8800	**4.** 1200	**4.** 570	**4.** 2665
5. 5358	**5.** 350	**5.** 1176	**5.** 2035	**5.** 2356	**5.** 6468
6. 2385	**6.** 812	**6.** 5841	**6.** 690	**6.** 1183	**6.** 3225

1. 3901	**1.** 1320	**1.** 765	**1.** 2989	**1.** 352	
2. 1764	**2.** 2112	**2.** 5720	**2.** 3304	**2.** 2849	
3. 1333	**3.** 1849	**3.** 6370	**3.** 5658	**3.** 2516	
4. 1760	**4.** 2665	**4.** 3003	**4.** 1240	**4.** 2200	
5. 2976	**5.** 480	**5.** 4214	**5.** 1488	**5.** 8370	
6. 5688	**6.** 520	**6.** 1221	**6.** 4158	**6.** 3825	

1. 29	**1.** 28	**1.** 46	**1.** 11	**1.** 26	**1.** 28
2. 41	**2.** 43	**2.** 38	**2.** 10	**2.** 23	**2.** 15
3. 40	**3.** 33	**3.** 14	**3.** 20	**3.** 48	**3.** 43
4. 24	**4.** 50	**4.** 22	**4.** 28	**4.** 10	**4.** 33
5. 13	**5.** 25	**5.** 45	**5.** 28	**5.** 10	**5.** 35
6. 45	**6.** 11	**6.** 32	**6.** 23	**6.** 32	**6.** 49

1. 14	**1.** 30	**1.** 50	**1.** 45	**1.** 33	
2. 26	**2.** 28	**2.** 11	**2.** 21	**2.** 26	
3. 18	**3.** 11	**3.** 34	**3.** 34	**3.** 31	
4. 18	**4.** 28	**4.** 14	**4.** 21	**4.** 40	
5. 31	**5.** 30	**5.** 44	**5.** 22	**5.** 32	
6. 26	**6.** 16	**6.** 20	**6.** 50	**6.** 13	

1. 4/9	**1.** 2/81	**1.** 7/50	**1.** 1 23/40	**1.** 8/25	**1.** 40/63
2. 1/2	**2.** 4/15	**2.** 1/2	**2.** 1/4	**2.** 2 2/3	**2.** 4
3. 7/50	**3.** 3/20	**3.** 4/9	**3.** 1/3	**3.** 5	**3.** 1 1/2
4. 3/10	**4.** 1/20	**4.** 4/5	**4.** 2 1/3	**4.** 1/2	**4.** 3/7

1. 14/27	**1.** 20/81	**1.** 3/50	**1.** 3	**1.** 8/9
2. 1/10	**2.** 32/81	**2.** 3/32	**2.** 9	**2.** 1 1/2
3. 15/64	**3.** 14/27	**3.** 1/4	**3.** 12/35	**3.** 16/25
4. 21/32	**4.** 3/10	**4.** 3/32	**4.** 14/15	**4.** 1/6

Made in the USA
Middletown, DE
18 January 2018